9/97

From Mare's Tails to Thunderheads **CLOUDS**

by Suzanne Harper

A First Book

FRANKLIN WATTS

A DIVISION OF GROLIER PUBLISHING

New York ■ London ■ Hong Kong ■ Sydney

Danbury, Connecticut

For my parents,
who taught me to look at the sky

Photographs ©: Archive Photos: 6 (Lambert); Brown Brothers: 37; Comstock: 24 (Art Gingert); Hughes Aircraft Company, Space and Communications Group, Public Relations Department: 45; John Morgan: 3; Kent Wood: cover, 9, 12, 14, 15, 16, 17, 18, 19, 22 top, 23, 26; Photo Researchers: 34 (Howard Bluestein), 47 (Mark C. Burnett), 50 (KCET TV/SS), 22 bottom (Keith Kent/SPL), 49 (NASA/SS); UPI/Corbis-Bettmann: 38, 40, 42.

Library of Congress Cataloging-in-Publication Data

Harper, Suzanne.
Clouds: from mare's tails to thunderheads / by Suzanne Harper
p. cm. — (A First book)
Includes bibliographical references and index.
Summary: Discusses cloud types, cloud formations, and how clouds affect the climate on earth and other planets. Includes experiments.
ISBN 0-531-20291-7
1. Clouds—Juvenile literature. 2. Cloud physics—Juvenile literature.
[1. Clouds.] I. Title. II. Series.
QC921.35.H37 1997
551.57'6—dc20

96-33307
CIP
AC

CONTENTS

INTRODUCTION

In ancient times, people believed that gods created the weather. When drought threatened their food supply, they held rainmaking ceremonies. These ceremonies sometimes involved animal or even human sacrifice.

Eventually, people learned to predict the weather by studying the sky and the clouds. They knew that certain kinds of clouds led to certain kinds of weather, and came up with sayings that have become part of our weather folklore.

At one time, sailors predicted weather using the folk saying, "Red sky at night, sailor's delight; red sky in morning, sailors take warning."

For example, sailors used to say, "Red sky at night, sailor's delight; red sky in morning, sailors take warning." This saying holds true because weather patterns usually move from west to east. A red sky at dawn means that the sun is reflecting off storm clouds that are moving from the west to the east (where the sun is rising).

Another folk saying warned, "Mare's tails and mackerel scales make tall ships take in their sails." A mare's tail is a cloud that looks like the wispy tail of a horse; a mackerel sky has clouds that look like fish scales. Mare's tails usually appear before a cold front arrives, while a mackerel sky usually appears before a warm front moves into an area. The arrival of either a warm front or a cold front is often associated with stormy weather.

In the eighth century B.C., philosophers in Asia Minor developed the first theories about how clouds form. They believed that clouds are composed of moist air that has thickened.

Centuries later, in the mid-1600s, the French mathematician René Descartes realized that air and water vapor are two different substances. Then in 1751, a French doctor named Charles Le Roy conducted an experiment. He sealed damp air in a glass container and let it cool. When the air reached a certain temperature, dew appeared on the inside of the

Mare's tails, which are actually a type of cirrus cloud, usually appear before a warm front or cold front arrives. As a result, they warn of stormy weather.

glass. When Le Roy heated the glass above this temperature, the dew disappeared. The results of this experiment showed that at a specific temperature, called the *dew point*, the water vapor in air condenses and becomes liquid water.

As scientists continued to study clouds, they began to develop scientific theories that explained why many of the old folk sayings could often accurately predict weather patterns. Today, *meteorologists*—scientists who study the weather—know what types of *atmospheric* conditions lead to various types of weather. They use satellites and other scientific equipment to track storms and predict weather patterns.

TYPES OF CLOUDS

CHAPTER 1

Although people observed the relationships between clouds and weather for centuries, no one classified and named clouds until the early 1800s. In 1803, an English pharmacist named Luke Howard wrote a paper called "On the Modifications of Clouds." Historians believe that Howard may have first become interested in clouds in 1783, when the eruption of a Japanese volcano caused amazing cloud formations around the world. Even though Howard was only an amateur meteorologist, his simple classification system is still used today.

Howard's system divided clouds into three basic categories based on their shape. He called clouds that are layered *stratus*, from a Latin word meaning

"stretched out"; clouds that look like large heaps *cumulus*, from the Latin for "heap" or "pile"; and clouds that are feathery *cirrus*, from the Latin for "curl" or "tuft."

Scientists attach additional terms to Howard's names when they want to describe clouds more precisely. For example, *alto*—meaning high—is used to describe certain clouds in the middle layer of the atmosphere. The word *fracto* refers to clouds that have been broken apart by the wind. The word nimbus is included in the names of clouds that produce *precipitation* (rain, hail, snow, or sleet). Thus, a cumulonimbus cloud is a puffy cloud with rain falling from it.

Today, meteorologists also classify clouds by their distance above the earth. *High clouds* include clouds whose bases are an average of 20,000 feet (6,100 m) above the earth. High clouds include cirrus, cirrocumulus, and cirrostratus clouds.

Cirrus clouds, which look like thin, wispy strands, occur 26,000 to 42,000 feet (7,900 to 12,800 m)

In 1803, an English pharmacist named Luke Howard developed a simple classification system for clouds that is still used today. Howard called the clouds shown in this photo cumulus.

High clouds, like these cirrocumulus clouds (top) and cirrostratus (below), have bases that are an average of 20,000 feet (6,100 m) above the earth.

above the earth. Cirrocumulus clouds are thin and patchy; they often look ripped. They form at 20,000 to 25,000 feet (6,100 to 7,600 m).

Cirrostratus clouds are thin and sheet-like. They also form at about 20,000 to 25,000 feet (6,100 to 7,600 m). Cirrostratus clouds sometimes create a "halo" around the sun or moon.

Altocumulus clouds are puffy and either gray or white.

Altostratus clouds look like dense blue or gray sheets and often completely cover the sky.

Vertical clouds are cumulus clouds created by strong vertical air currents. They can form at almost any altitude. They include cumulus and cumulonimbus clouds. Cumulus clouds are puffy and look a little like cauliflower. These clouds change shape constantly and usually indicate fair weather. Although the bases of cumulonimbus clouds, also called thunderheads, almost touch the ground, the tops may be 75,000 feet (23,000 m) tall. The tops of these clouds often spread out, creating an anvil shape. They can cause violent weather, including tornadoes.

Vertical clouds, like these cumulus (top) and cumulonimbus clouds (right), are created by strong vertical air currents and can form at almost any altitude.

Nimbostratus clouds are dark and cause rain. Stratocumulus clouds often look layered.

Low clouds generally occur less than 6,500 feet (2,000 m) above the earth's surface. They include stratus, nimbostratus, and stratocumulus clouds. The base of a stratus cloud is close to the ground. Although stratus clouds are usually less than 3,200 feet (1,000 m) thick, a single stratus cloud may be wide enough to cover the states of Montana *and* Wyoming. Fog is a stratus cloud that's lying on the ground. Nimbostratus clouds, which cause rain, often appear darker than other stratus clouds. Stratocumulus clouds are dark gray, layered, and puffy.

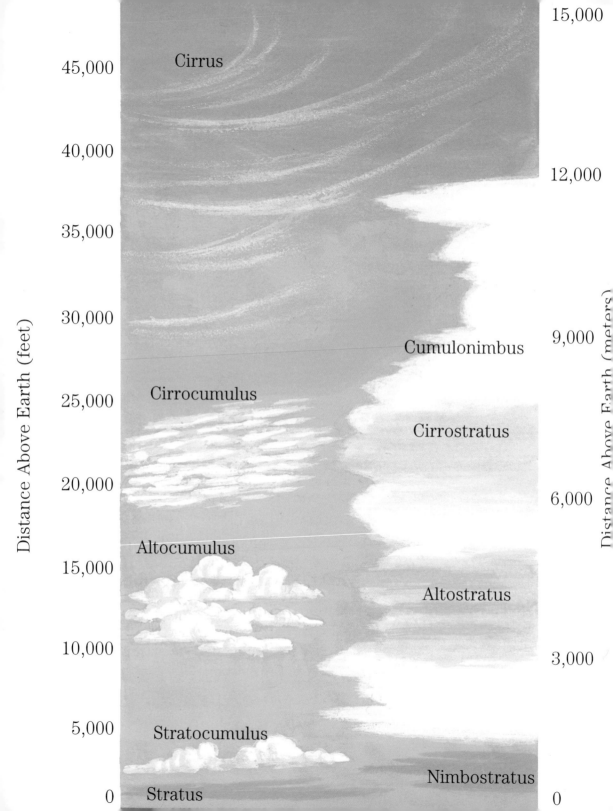

Bizarre Clouds

Some clouds have very unusual shapes. For example, people sometimes mistake *lenticular clouds* (also known as wave clouds), which look like flat disks, for alien spacecraft. These clouds most often appear downwind from hills or mountains. To picture how these clouds form, imagine that air flows like the water in a stream. When the water in a stream moves past a rock or branch, eddies or ripples appear in the water. In a similar way, wave clouds occur when rapidly moving air encounters slow-moving or stable air and a ground feature (such as a hill or mountain) creates eddies or turbulence in the air.

Sometimes cirrus clouds look bluish-silver and appear to brighten as the sky darkens at dusk. This effect is the result of the relative positions of the setting sun and the clouds. Even after the sun has moved below the horizon, some of its light can illuminate tiny dust particles in clouds. These clouds, which are called *noctilucent clouds*, seem to glow in the dark.

Have you ever noticed a thundercloud that has rounded bumps on its underside? When updrafts carry water droplets up to the top of the cloud, the water freezes and forms ice crystals. These ice crystals are so heavy that they fall back down through the cloud, pulling cool air down with them. At the same time,

Lenticular or wave clouds look like flat disks and are sometimes mistaken for alien spacecraft.

Noctilucent clouds appear to brighten as night falls.

*The rounded bumps on the bottom of
this cloud are called cloud pouches.
They form when a tower of rising warm air
pushes against pockets of falling cool air.*

warm air is rising inside the cloud. The rounded bumps
are cloud pouches that form as the tower of rising
warm air pushes against pockets of falling cool air.

When the sun or moon shines through clouds, you
can sometimes see a disk of colored light, called a

When the sun or moon shines through clouds, you can sometimes see a disk of colored light, called a cloud corona.

cloud *corona*. This disk occurs when the cloud bends the light from the sun or moon as it passes through. You shouldn't try to look at the sun, since it can damage your eyes and even blind you. However, if you look at the moon through the clouds, you can sometimes see that the corona is blue on the inside (nearest the moon) and reddish-brown on the outside.

Watching Clouds to Predict Weather

If we are wondering what tomorrow's weather will be, all we have to do is turn on a radio or television. By using radar and satellite images, meteorologists can provide us with fairly accurate information. Less than 100 years ago, however, people had no satellites; no television; and no radio. They had to learn how to predict weather patterns by watching the sky.

By learning to identify cloud types, you can predict weather, too. For example, rain or snow is possible when you see a layer of rippled high clouds (cirrocumulus); a milky white layer of high clouds that covers the sky (cirrostratus); middle clouds that are white or light-gray and have flattened layers (altocumulus); or a thick, smooth blanket of gray clouds that make the sun look dim (altostratus).

Cumulus clouds look like giant cotton balls and are usually a sign of fair weather. If cumulus clouds change into cumulonimbus clouds, a storm is on its way. Sometimes cumulus clouds grow so tall that they look like giant towers. These clouds are called altocumulus castellanus. If you see them in the morning, rain is possible in the afternoon or evening.

If the wind is from the southwest or the north, high, wispy cirrus clouds indicate good weather is on its way. If the wind is from another direction, these clouds will thicken and rain or snow is likely.

When cumulus clouds grow so tall that they look like giant towers, they're called altocumulus castellanus.

You may have seen an airplane fly across the sky, followed by long cloud trails. These *contrails* are cirrus clouds created when the water vapor from the plane's engines freezes into ice crystals. They appear when the plane flies from 25,000 to 40,000 feet (7,600 to 12,400 m) above the earth. Since contrails form only when the air contains moisture, they can help you predict the weather. If the contrail vanishes quickly, the weather will be fair; if it stays in the sky, a change in the weather may be coming.

THE BIRTH OF CLOUDS

CHAPTER 2

Have you ever spent a lazy hour or two lying on the grass and looking at the sky, trying to see shapes in the clouds? Did you see a cloud that was shaped like a familiar animal or object? If so, chances are you could see that shape for only a few minutes because clouds are constantly moving and changing.

At any moment in time, one-half of Earth's surface is covered with clouds. As you will see, the constant movement of these clouds provides us with a visible clue of what's happening in the atmosphere.

Clouds are born, in a sense, on the ground. Because the sun warms the earth unevenly, some parts of the ground are always warmer than others. When air comes into contact with a patch of warm earth, the air

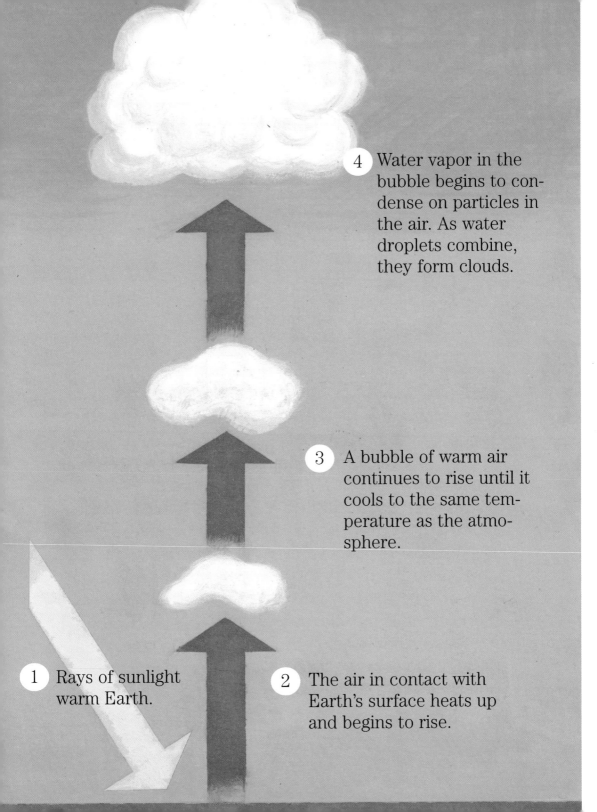

4 Water vapor in the bubble begins to condense on particles in the air. As water droplets combine, they form clouds.

3 A bubble of warm air continues to rise until it cools to the same temperature as the atmosphere.

1 Rays of sunlight warm Earth.

2 The air in contact with Earth's surface heats up and begins to rise.

warms up. As the air heats up, it begins to rise and expand, usually in the form of a bubble. It floats up in the atmosphere just like a hot-air balloon.

As the bubble—also called a *convection cell*—drifts upward, it begins to cool off. For every 1,000 feet (300 m) that the air rises, the temperature drops 5.4°F (2.8°C). As long as the surrounding air is cooler than the convection cell, the cell will keep rising and cooling.

Because cool air cannot hold as much water vapor as warm air, the water vapor in the bubble begins to condense on tiny particles in the air. These particles could be salt from the ocean, pollen, bits of rock, volcanic dust, pollution, or soot. As more and more *molecules* of water attach themselves to the particles, water droplets or ice crystals form. Those water droplets or ice crystals create what we see as clouds.

As the warm air in a convection cell rises, it comes into contact with cooler and cooler air. Eventually, the bubble breaks up and the air begins to sink. At any moment in time, half of the air in the atmosphere is rising, and the other half is sinking. As warm air rises into the atmosphere, it is replaced by cooler air. The cooler air that comes into contact with the Earth cools it. Thus, clouds play an important role in heating and cooling our planet.

Even though a convection cell usually breaks up in about 20 minutes, the cloud it has formed may last for an hour or more because new cells bubble up from the sun-warmed earth. If there is a wind, the bubbles may drift as they rise, creating lines of clouds for miles.

If the air surrounding a cloud is dry, the cloud will evaporate and disappear. However, if the surrounding air holds large quantities of moisture, the cloud evaporates much more slowly.

Tall mountains may block moist air currents and push them higher into the atmosphere, causing clouds to form. These clouds rain heavily on the windward side of the mountains. Because moist air currents rarely make it over the mountains, deserts often form on the opposite side.

Why Clouds Float

In the nineteenth century, scientists puzzled over how clouds manage to stay up in the air at all. Water is obviously heavy, they thought. So if clouds are made of water, how can they float? Some scientists believed the water drops in clouds actually resemble hot-air balloons—bubbles with fires inside them that would keep them aloft.

Scientists now know that the force of rising air keeps cloud droplets aloft. This is amazing when you consider how much a cloud can weigh. Although it looks fluffy and light, a small cumulus cloud may weigh about 550 tons—more than an airplane carrying a full load of passengers.

Why Clouds Are White

The light coming from the sun is white. When this light hits a drop of water, the water drop acts like a prism—it breaks the light up into the colors of the rainbow: red, orange, yellow, green, blue, and violet.

So if sunlight is white and air has no color, why do we use a yellow crayon to draw the sun and a blue crayon to draw the sky? The sky appears blue because the air and water molecules in it scatter the colors blue and violet in all directions, while red, orange, and yellow light from the sun pass through straight to our eyes. When you look at the sun in a photograph, it looks yellow because you are actually seeing the light rays that pass straight through the atmosphere. ***Never look directly at the sun. It's light is so bright that it may damage your eyes.***

When sunlight hits the water molecules in a cloud, all the colors scatter and we begin to see all the colors recombined which, to our eyes, looks white. This is why most clouds look white. Storm clouds often look dark-gray because they are so thick that most of the sun's light cannot pass through them.

PRECIPITATION

CHAPTER 3

C louds are fun to watch but, as everyone knows, they may bring nasty weather— rain, hail, sleet, or snow. Actually, only 10 percent of clouds produce precipitation.

As long as moist, warm air continues to rise, its upward motion prevents water molecules from combining to form water droplets. Even when the updrafts stop, the water droplets that form are so small that they evaporate before they reach the ground. Before a cloud begins to precipitate, many, many water droplets must combine. It takes about a million of these water droplets to form a rain drop.

If the temperature of the air between a cloud and the ground is less than 32°F (0°C), the water droplets turn into snow or hail as they fall. If the

*Although most hailstones are less than
1 inch (2.5 cm) in diameter, some are the size
of golf balls or even baseballs.*

temperature is a little warmer, the water droplets form sleet—a combination of rain, snow, and partly melted snowflakes.

Sometimes, precipitation starts to fall, and is lifted up again by rising air. A raindrop, snowflake, or hailstone may fall out of one updraft and then be carried upward by another updraft. In some cases, the precipitation may fall and be lifted up many times. Each time it is lifted up, it may get bigger. A hailstone, for example, may develop another ice coating each time it's lifted up, resulting in a large hailstone that has many layers. While most hailstones are less than 1 inch (2.5 cm) in diameter, some may be larger than a baseball.

Trying to Make Rain

Long periods without rain are called droughts. These dry periods can kill crops and livestock, and reduce our reserves of drinking water. Because we are so dependent on rain, people have tried to figure out ways to make it rain on demand.

In ancient cultures, the rainmaker was one of the most important people in the community. It was his job to conduct rituals that would bring down the precious rain when it was needed. Sometimes the rituals involved the entire tribe. For example, the Nootka Indian tribe in British Columbia painted their faces black, then washed them. This ritual represented rain falling from black clouds.

In the late 1800s, settlers began farming the Great Plains and encountered severe droughts for the first time. Although many scientists didn't believe in rainmaking, most people were willing to try anything, even if it seemed farfetched.

One theory was that artillery fire caused rain. In 1891, the United States Congress actually spent more than $20,000 testing this idea. None of the tests were successful. Some people believed that smoke from prairie fires affected clouds and caused rain. Again, this was never proven.

During this time, men called "rainmakers" traveled throughout the country, promising to bring rain to parched farmland. Their methods ranged from special chemical mixtures to electricity.

One of the most famous rainmakers was Charles Mallory Hatfield. Although he had many successes, some skeptics claimed that Hatfield merely studied weather patterns and carried out his experiments at a time when rain would probably have arrived on its own.

In 1916, the city council of San Diego, California, hired Hatfield to break a drought. Hatfield and his brother built a 20-foot (6-m) tower near Morena Dam and cooked their chemical mixture over a fire. The

WEATHER CONTROL BUREAU
507 5TH AVE.
NEW YORK CITY. Dr G.A.I.M. SYKES.

*In the late 1800s, men called "rainmakers"
traveled all over the United States, promising
to bring rain to drought-ridden areas.*

next day, a heavy rainfall started at noon and continued for 7 days. Six days later it began to rain again.
This time the rain lasted for 3 days. During that
month, 38 inches (96 cm) of rain fell—a record for the
area that has never been broken.

All this rain was too much of a good thing. The
flood that resulted washed away bridges and railroad
tracks, destroyed private property, and caused fifteen

deaths. Hatfield was almost lynched by angry citizens who blamed him for the flood damage.

The flood generated so much publicity that Hatfield was hired by a number of other towns desperate for rain. Although no one knows what chemi-

In 1916, Charles M. Hatfield, one of the most famous rainmakers, was hired to break a drought in San Diego, California. So much rain fell that a flood occurred. The flood destroyed property and caused fifteen deaths.

cal formula Hatfield used to create rain, he did tell people that he mixed twenty-three chemicals, let them age for a few days in casks, and then put them in evaporating pans on top of his towers.

Cloud Seeding

Although rainmaking is often viewed as more of a con game than a science, researchers are still investigating ways to "seed" clouds to create rain. During World War II, an American scientist named Vincent Schaefer began studying the rime ice that had formed on objects on high mountains. He also studied various types of surfaces that prevent ice from forming, even when the temperature is below freezing. He hoped his experiments would lead to ways of preventing the buildup of ice on military planes flying at high altitudes.

Schaefer put dry ice (frozen carbon dioxide) in a chamber containing a fog of supercooled water droplets—water droplets that had been cooled to below the freezing point, but had not turned to ice. When the carbon dioxide was added, the fog instantly turned into a snowstorm. To see if he could create a snowstorm in the atmosphere, Schaefer flew over an altocumulus cloud and dropped 3 pounds (1.4 kg) of dry ice into the cloud. As he did this, people on the ground saw snow falling from the cloud.

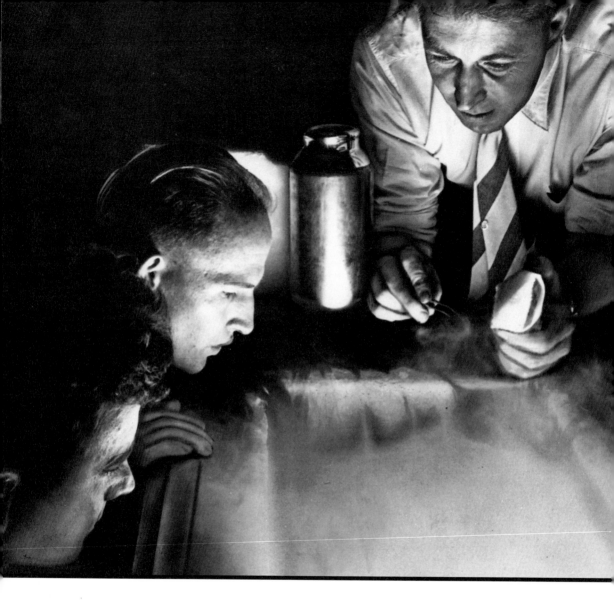

Scientist Vincent Schaefer (top right) developed the most popular method of cloud seeding. Here, he waves an ice wand through water vapor to create water droplets.

Later, one of Schaefer's co-workers, Bernard Vonnegut, tried sprinkling clouds with a chemical called silver iodide. Vonnegut was able to make rain come down from the clouds. Silver iodide is still the chemical most often used to squeeze more rain from clouds.

Cumulus clouds can be seeded in one of two ways. Airplanes can fly over a cloud and drop a small explosive charge that releases silver iodide particles. Alternatively, the chemical can be released below the cloud, either from a plane or from a generator on the ground, and carried into the cloud on an updraft of air.

Experiments conducted by scientists in Israel have suggested that seeding a cloud with silver iodide may increase rainfall by about 15 percent in the target areas, but other scientists have not produced such clearcut results. That is the main problem with cloud seeding—no one is sure it really works because no one knows whether rain or snow would have fallen at that time even if the cloud seeding had not taken place.

While there is no real proof that seeding works, many people believe that it does. Recently, farmers in Montana claimed that clouds seeded over the eastern part of their state actually rained on North Dakota. They insisted that North Dakota farmers had reaped

During a 1949 drought, a reporter tried seeding moisture-laden clouds with dry ice to create rain. The main problem with cloud seeding is that no one knows if it really works—
or if rain would have fallen naturally.

the benefits of cloud seeding paid for by the state of Montana. After a legal battle, the state of North Dakota agreed to pay for research to show that farmers in Montana had received the rain they paid for.

Unfortunately, cloud seeding can be used only with thick cumulus clouds that are about to produce rain naturally. In Chile, scientists are trying to develop ways of "catching clouds." They set up nylon mesh nets that are 39 feet (12 m) long and 13 feet (4 m) wide on hilltops. When fogs called camanchacas blow through the nets, the mesh strains out drops of water. This process yields about 2,500 gallons (9,500 l) of water each day.

SCIENTISTS STUDY THE CLOUDS

CHAPTER 4

People have been watching the skies—and studying clouds—for centuries. Today, scientists collect information about clouds from weather satellites that orbit the Earth and from planes that fly through the clouds to make measurements and gather ice crystal samples.

Scientists are also performing experiments to find out more about the size and shape of the ice crystals that make up clouds. They measure the amount of sunlight reflected by cloud particles and the amount of infrared energy (or heat) they give off; the motion of air inside the clouds; and the relative amount of water, water vapor, and ice inside clouds. They also study how clouds affect large-scale weather patterns.

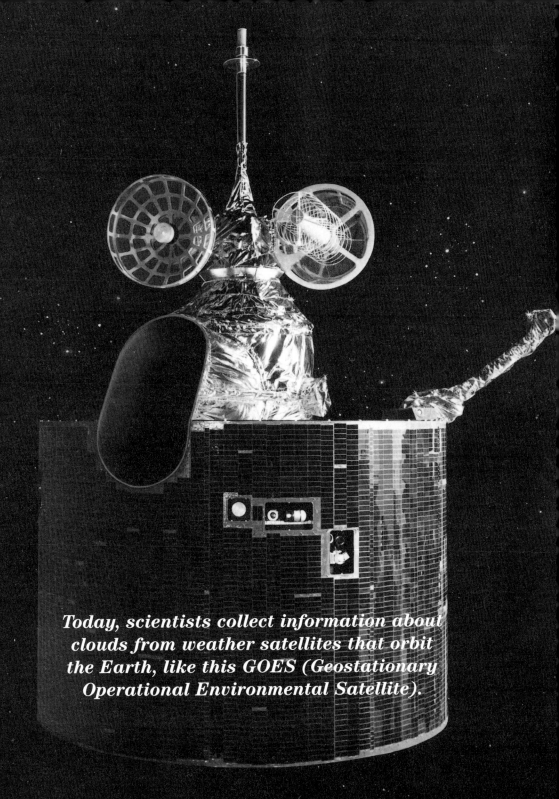

Today, scientists collect information about clouds from weather satellites that orbit the Earth, like this GOES (Geostationary Operational Environmental Satellite).

A wide variety of equipment is used to gather information about clouds and atmospheric conditions. *Lidar*, or laser light radar, and *Doppler radar* measure cloud height, structure, and movement. Lidar can also measure the size and shape of particles that make up a cloud.

Wind profilers measure the speed and direction of the wind; *radio acoustic sounders* use sound waves to measure upper-air temperatures; and *electromagnetic sensors* measure infrared energy and light. Scientists also send a variety of instruments up into the clouds in weather *balloons*.

Clouds and Our Climate

Scientists want to know as much as they can about clouds because Earth's climate seems to be controlled by clouds. Researchers believe that clouds help regulate the planet's temperature. They cool the planet by reflecting sunlight away. They also warm the planet by absorbing infrared radiation from the sun, and then radiating some of it back down to Earth.

As you learned in Chapter 3, when the earth heats up, water evaporates, and the water vapor rises into the atmosphere. The hotter the earth grows, the more water evaporates, and the more water vapor is released into the atmosphere. This

Scientists use weather balloons to carry a variety of weather instruments into the clouds.

water vapor forms thick, low-altitude clouds. These thick clouds can reflect sunlight back into space, and, as a result, the earth cools. Some scientists estimate that, without clouds, the Earth might be as much as 20°F (11°C) hotter.

When we pollute the atmosphere, we alter this cycle. Air pollution may cause chemical reactions in clouds. These reactions may result in smaller water droplets, which scatter more light. If the tops of clouds around the world are becoming brighter and reflecting more light into space, it is possible that air pollution is cooling the earth.

Clouds on Other Planets

Earth is not the only planet that has clouds. Venus, the second planet from the sun, has an atmosphere made up almost entirely of carbon dioxide. The thick layer of yellow clouds around Venus is made up of concentrated sulfuric acid. These clouds block out most of the sun's rays, so a typical day on Venus is no brighter than a very cloudy day on Earth.

Scientists have also observed cloudlike formations on Mars, the fourth planet from the sun. Many of these observations were made by the six unpiloted spacecraft launched by the United States between 1964 and 1976. Bright clouds, believed to consist of carbon dioxide, form over the planet's

Venus, the second planet from the sun, is surrounded by a thick layer of yellow clouds made up of concentrated sulfuric acid.

polar regions during its autumn. Despite the slight amount of water vapor present in Mars's atmosphere, scientists have occasionally observed hazes at high altitudes and even clouds composed of water and ice. These cirrus clouds often have unusual and intriguing shapes.

*Frozen ammonia forms white cirrus clouds
in the upper atmosphere of Jupiter,
the fifth planet from the sun.*

Strong winds sometimes stir up the sand on Mars's surface and cause dust to be suspended in the atmosphere. These dust storms are occasionally large enough to cover the planet's surface for weeks—or months—at a time.

Jupiter is the fifth planet from the sun and the largest planet in our solar system. Its cold atmosphere is filled with clouds. The planet's upper atmosphere reaches −193°F (−125°C), a temperature cold enough to freeze the ammonia in its atmosphere. This frozen ammonia forms white cirrus clouds. Lower in Jupiter's atmosphere, the clouds are often colored by chemical compounds, resulting in a tawny cloud layer and a vast oval area called the Great Red Spot. Jupiter's clouds have also been photographed from unpiloted spacecraft, revealing changes that seem to indicate the birth and death of giant storm systems on the planet.

Saturn, the sixth planet from the sun and the second largest planet in our solar system, has whirls and eddies of clouds in its atmosphere. These clouds were photographed by *Voyager 1* and *Voyager 2*. When scientists looked closely at these photographs, they saw what appeared to be a giant hexagon near Saturn's north pole. This hexagon was caused by cloud bands around the pole.

Planets are not the only astronomical bodies with clouds. There are even extremely hot hydrogen clouds above the surface of the sun.

EXPERIMENTS YOU CAN DO

CHAPTER 5

Would you like to make your own cloud or see how rain really forms? Would you like to learn how to predict the weather yourself? The experiments on the next few pages are a good place to start.

Experiment 1–Creating a Cloud

You will need an empty glass bottle (such as a 16-ounce [0.5 l] soda bottle), matches, and a candle. **Before performing this experiment, ask an adult for assistance. Working with fire can be dangerous.**

First, light the candle. Then turn the bottle upside down and hold the lit candle inside the mouth

Bottle

Matches

Candle

A cloud forms when the water vapor from your warm breath condenses on soot from the candle.

of the bottle for 5 seconds. When the bottle has cooled, put your mouth over the opening and blow hard into the bottle. Now take your mouth away and watch what happens. You should see a small cloud form inside the bottle.

Why does this cloud form? A cloud requires two ingredients: 1) tiny particles of soot, dust, or pollen and 2) moist air. The soot was supplied by the candle as it burned and your breath provided the moist air. When you took your mouth away from the bottle, the air inside the bottle cooled. The water vapor condensed, and then turned back into liquid on the soot particles. The result: tiny water droplets—like the water droplets that form a cloud in the sky.

Experiment 2–Making Rain

You will need a kettle, a stove, a can filled with ice, and ice tongs. **Before performing this experiment, ask an adult for assistance. A hot stove can be dangerous.**

Pour water in the kettle until it is half full and then put it on the stove to boil. When steam starts rising from the kettle, hold the ice-filled can above the steam with the tongs, and watch what happens. You should see a few drops of "rain" fall from the bottom of the can.

Ice tongs

Can with ice

Tea kettle

As steam from the tea kettle melts the ice inside the can, water vapor forms on the outside of the can. The water that drops off the bottom of the can is "rain."

This experiment basically re-creates the way rain is formed in nature. The sun warms the water in lakes, rivers, and oceans. The water evaporates and becomes water vapor, which rises into the air. As the water vapor rises in the atmosphere, it cools and condenses into rain. In this experiment, the stove acts like the sun. As you heat the water in the kettle, some of the water evaporates and forms steam. When the steam comes into contact with the ice-filled can, it condenses and forms "rain."

Experiment 3—Keep Your Eyes on the Skies!

You will need a chart of different cloud types and a journal (a notebook). You may be able to find a chart at a local bookstore or library. Begin by going outside in the morning. Look up at the sky. What type of clouds do you see? In your journal, write down the cloud types and the current weather conditions. About 6 hours later, go back outside and see what type of clouds are in the sky. Again write down the type of clouds you see and the weather conditions. Repeat this procedure for several days. Do you notice any pattern between the type of clouds you see in the morning and the weather conditions in the afternoon? You should. If you keep doing this, you can learn to predict the weather.

GLOSSARY

alto—a prefix that means high; it refers to clouds that form in the middle layer of the atmosphere.

atmosphere/atmospheric—the mass of air that surrounds the earth.

cirrus cloud—a cloud that looks feathery (from the Latin word for "curl" or "tuft").

contrail—a cirrus cloud created when the water vapor from an airplane's engine freezes into ice crystals.

convection cell—a bubble of warm air that rises from the earth and creates a cloud.

corona—a white or colored circle of light seen around the sun or the moon.

cumulus cloud—a cloud that forms in heaps (from the Latin word for "heap" or "pile").

dew—small droplets of water.

dew point—the specific temperature at which water vapor in the air condenses to liquid water.

Doppler radar—a scientific instrument that measures cloud height, structure, and movement.

drought—a period of dryness.

electromagnetic sensor—a scientific instrument that measures infrared energy and light.

fracto—a prefix that refers to clouds that have been blown apart by the wind.

high cloud—a cloud whose base is an average of 20,000 feet (61,000 m) above the earth.

lenticular cloud—a cloud that looks like a flat disk (the word comes from "lens- shaped").

lidar—a device similar to radar that uses pulsed laser light to measure the height, structure, and movement of clouds as well as the size and shape of particles that make up clouds.

low cloud—a cloud found up to 6,500 feet (2,000 m) above the earth's surface.

meteorologist—a scientist who studies the atmosphere, weather, and climate.

middle cloud—a cloud that's found about 10,000 feet (3,050 m) above the earth.

molecule—a group of atoms that forms the smallest unit of a substance that can exist and retain its chemical properties.

nimbus cloud—a cloud that produces precipitation (rain, hail, snow, or sleet).

noctilucent cloud—a cloud that seems to glow in the dark.

precipitation—the falling product of condensation in the atmosphere, such as rain, hail, or snow.

radioacoustic sounder—a scientific instrument that uses sound waves to measure upper-air temperatures.

stratus cloud—a cloud that forms in layers (from a Latin word meaning "stretched out").

vertical cloud—a cumulus cloud that is created by strong vertical air currents and that can form at almost any altitude.

weather balloon—a balloon used to carry scientific instruments into the clouds.

wind profiler—a scientific instrument that measures the speed and direction of the wind.

ADDITIONAL INFORMATION

Books

Cosgrove, Brian. *Weather*. New York: Alfred A. Knopf, 1991.

Simon, Seymour. *Storms*. New York: William Morrow and Company, 1989.

Trefil, James. *Meditations at Sunset: A Scientist Looks at the Sky*. New York: Macmillan, 1987.

Wagner, Ronald L. and Bill Adler, Jr. *The Weather Sourcebook: Your One-Stop Resource for Everything You Need to Feed Your Weather Habit*. Old Saybrook, CT: Adler & Robin Books, Inc., 1994.

Watson, Benjamin A. and the editors of *The Old Farmer's Almanac. The Old Farmer's Almanac Book of Weather and Natural Disasters*. New York: Random House, 1993.

Williams, Jack. *The Weather Book: An Easy-to-Understand Guide to the USA's Weather.* New York: Vintage, 1992.

Wind and Weather. New York: Scholastic, 1994.

Wyatt, Valerie. *Weatherwatch*. New York: Addison-Wesley, 1990.

Magazine Articles

Baskin, Yvonne. "Under the Influence of Clouds." *Discover*. September 1995, pp. 62–69.

Jesiolowski, Jill. "Look! Up in the Sky! Learn to Predict Your Own Weather!" *Organic Gardening*. December 1993, pp. 24–30.

Maxham, Glen. "Spirit of the Wind." *Weatherwise*. October/November 1991, p. 9.

CD-ROM Articles

"Jupiter," "Mars," "Saturn," "Venus," Microsoft® Encarta® 97 Encyclopedia, Deluxe Edition. ©1993–1996 Microsoft Corporation. Funk & Wagnall's Corporation.

Internet Resources

Due to the changeable nature of the Internet, sites appear and disappear very quickly. These resources offered useful information on clouds at the time of publication.

The National Climatic Data Center publishes satellite photos of clouds and storms on its Web site. It's located at **http://www.ncdc.noaa.gov/ncdc.html.**

The Cloud Catalog site contains images and information compiled by the Department of Atmospheric Sciences at the University of Illinois at Urbana-Champaign. It's located at **http://covis.atmos.uivc.edu/guide/clouds.**

The Cloud Gallery site contains photographs of various types of clouds. It's located at **http://www.commerce.digital.com/ paloalto/CloudGallery/home.htm.**

INDEX

Page numbers in *italics* indicate illustrations.

ABOUT THE
AUTHOR

Suzanne Harper is the executive editor of *Disney Adventures* and has published numerous magazine and newspaper articles. She earned journalism and English degrees from the University of Texas-Austin and a master's degree in writing from the University of Southern California. She lives in New York City.